Introduction to Geology: India's Geological Wonders

By Siva Prasad Bose

Contents

Dedication

This book is dedicated to the sacred and majestic land of India, also called Bharat.

Preface

———

Delhi shakes, tremors ripple through Uttarakhand, and the towering Himalayas inch skyward each year. What causes this hidden drama beneath our feet? From the massive, flat Indo-Gangetic plains to the rugged peaks of the Himalayas, India's landscape is alive with geological forces. This book dives into the incredible forces shaping our country, including earthquakes, ancient land movements, and deep-earth mysteries.

We often take our landscapes for granted, but every mountain, valley, and coastline tells a fascinating tale. This book makes geology come alive. It covers things like why earthquakes rattle Delhi, why the Deccan Plateau has rich red soil, and why Sri Lanka looks like a missing jigsaw piece of Kerala. Through engaging stories and surprising facts, we explore India's deep geological history in a way anyone can understand.

This book is divided into the following themes:

- A Shaking Land: Why Delhi and the North experience earthquakes.
- The Making of India: The rise of the Himalayas, the mystery of the vast Indo-Gangetic plain, and the breakup of Gondwana.
- The Fiery Past: Deccan volcanism, its long-lasting impact, and the secrets behind its red soil.
- Coasts, Islands, and Forgotten Lands: How Sri Lanka and Kerala share a geological past, the creation of the Andaman and Nicobar Islands, and the story of submerged lands.
- How It Affects Us: Understanding India's earthquakes, mineral wealth, and the impact of geology on our daily lives.

Acknowledgements

In preparing this book, the authors would like to acknowledge various sources including the following books related to geology:

- Structural geology by Marland P. Billings
- Introduction To Geology by Dr. Vinod Agrawal, Dr. Harish Kapasya, Dr. Hemant Sen
- Principles of Engineering Geology by K.M. Bangar
- Physical geology: Mallory, B. & D. Cargo
- Geology of India and Burma by M. S. Krishnan
- Geological Survey of India publications and technical reports (www.gsi.gov.in)
- National Center for Seismology, Ministry of Earth Sciences, Government of India (seismo.gov.in)
- Wikipedia contributors, various articles on Indian geology, plate tectonics, and geological features of the Indian subcontinent

Chapter 0: How the Earth Works: A Simple Guide to the Forces Beneath Us

Before we explore India's mountains, plains, coastlines, and ancient lands, it helps to understand the basic machinery of our planet. Geology may seem like a story of giant forces and deep time, but its foundations are surprisingly easy to grasp. Every earthquake, volcano, river, and coastline is shaped by a few simple principles. This chapter introduces those ideas in clear language, so the rest of the book becomes easier to follow.

0.1 The Layers of Our Restless Planet

The Earth looks calm from the surface, yet beneath our feet lies a world of motion. Geologists divide the Earth into the following layers:

- **The Crust**: This is the thin, solid skin on which we live. Compared to the size of the planet, the crust is incredibly thin, like the peel of an apple. India sits on a piece of this crust called the *Indian Plate*.
- **The Mantle**: Below the crust is a thick layer of hot rock that flows slowly like very thick porridge. The mantle drives the movement of continents.
- **The Core**: At the centre lies the core, made mostly of iron and nickel. The inner core is solid; the outer core is molten and generates Earth's magnetic field.

The movement in the hot mantle beneath us is the engine that moves continents, lifts mountains, and creates earthquakes.

0.2 Plate Tectonics: Why Continents Move

The crust is broken into large pieces called *tectonic plates*. These plates float on the slowly moving mantle. Some plates collide, some drift apart, and some slide past each other. Their movement is slow, usually a few centimetres per year, but over millions of years it reshapes entire continents.

India sits on its own plate, the Indian Plate, which has been moving northward for more than 100 million years. This movement created the Himalayas, shaped the Indo-Gangetic Plain, and continues to cause earthquakes today.

Understanding plate tectonics is like understanding the grammar of geology. Every major landscape in India, from the Himalayas to the Deccan Plateau, is written by this grammar.

0.3 Volcanoes: Windows into the Deep Earth

Where plates pull apart or one plate sinks under another, magma from the mantle can rise and erupt as lava. This is how volcanoes form.

There are three main settings where volcanoes appear:

- **Subduction zones**, where one plate slides beneath another (like the Andaman region).
- **Rift zones**, where plates move apart.
- **Hotspots**, where plumes of molten rock rise from deep inside the Earth (such as the one that created the Deccan Traps).

Volcanoes are not simply destructive. They create new land, build mountains, enrich soils, and reveal the chemistry of Earth's interior.

0.4 Rocks: The Storytellers of Deep Time

Rocks hold the memory of the Earth. Geologists group them into the following three main types:

- **Igneous Rocks**: Formed when magma cools. Basalt in the Deccan Plateau is igneous.
- **Sedimentary Rocks**: Formed when sand, mud, shells, or organic material settle in layers. Fossils are mostly found in these rocks.
- **Metamorphic Rocks**: Formed when existing rocks change under heat and pressure. Marble and slate are examples.

These rocks constantly change from one form to another in a process known as the rock cycle. A rock can begin as lava, become part of a mountain, erode into a riverbed, and eventually turn back into rock again.

Understanding the rock cycle helps explain everything from the red soils of the Deccan to the soft alluvial soils of the Ganga plains.

0.5 Earthquakes: When the Planet Releases Energy

Earthquakes happen when rocks deep underground suddenly break or shift. This release of energy sends shockwaves through the ground. Most earthquakes occur along fault lines, which are breaks in the crust where plates meet or where old fractures still react to deep forces.

India experiences frequent earthquakes because of the following factors:

- The Indian Plate is still colliding with the Eurasian Plate.

- The Himalayas are a young and active mountain system.

- Old faults in central and southern India occasionally reactivate.

Earthquakes cannot be prevented, but understanding why they occur helps us build safer cities and prepare for future risks.

0.6 How Geologists Know the Age of the Earth

Many landscapes in India, such as the ancient rocks of Karnataka or the fossils of Rajasthan, are hundreds of millions or even billions of years old. Geologists determine these ages using some of the following techniques:

- **Radiometric Dating**: Measuring how radioactive elements inside minerals decay over time. This allows very precise dating of rocks.
- **Fossils**: Certain fossils appear only in specific time periods. Their presence helps date the rock layers they are found in.
- **Stratigraphy**: Studying the order and position of rock layers to understand the sequence of events.

These scientific tools allow us to reconstruct India's geological journey from the deep past of Gondwana to the creation of the Himalayas.

0.7 Slow Processes That Shape Our Lives

Many geological processes move slowly but have enormous long-term effects. Some of them are as follows:

- **Erosion**: Rivers, wind, glaciers, and waves wear down mountains and carry sediments far away.
- **Deposition**: Rivers drop these sediments to create plains, deltas, and fertile farmland.
- **Weathering**: Sunlight, rain, and temperature slowly break down rocks, forming soil.
- **Sea-level change**: Rising or falling seas reshape coastlines and drown ancient settlements.

These slow processes made the Indo-Gangetic Plain flat, carved India's coastlines, and shaped the Deccan's soils.

0.8 Why a Basic Understanding Matters

Geology may seem remote, but it shapes every part of our daily lives. Some of the ways are as follows:

- Why some cities are safer from earthquakes.
- Why some regions have fertile soil while others face drought.
- Why certain minerals, such as gold, coal, iron, diamonds, occur where they do.
- Why coastlines change and rivers shift.
- Why climate change threatens the Sundarbans and Himalayan glaciers.

Understanding these principles allows us to read the Indian landscape with clearer eyes. It prepares us for the more detailed chapters that follow, about rising mountains, vanishing coasts, ancient supercontinents, and the fiery origins of the Deccan Plateau.

Geology is the quiet force that has shaped India for billions of years. Learning its basic language helps us appreciate the country's deep past and prepare for its uncertain future.

Part 1: A Shaking Land – Earthquakes and Rising Mountains

Chapter 1: Why Delhi Trembles

Delhi sits on the edge of a restless zone. Why do tremors shake the city every few years? The answer lies beneath the Indo-Gangetic Plain, where the Indian plate grinds against the Himalayas. The region's fault lines store energy for centuries before releasing it as quakes. This chapter explores past earthquakes and what they mean for the future.

1.1 The Hidden Fault Lines Beneath Delhi

For millions of people living in and around Delhi, earthquakes seem like an occasional inconvenience, something to notice only when the ground shakes for a few seconds. But beneath the capital, immense geological forces are at work, setting the stage for future tremors. Delhi sits precariously close to two major seismic zones: the Himalayan seismic belt to the north and the Delhi-Moradabad fault system beneath the Indo-Gangetic Plain.

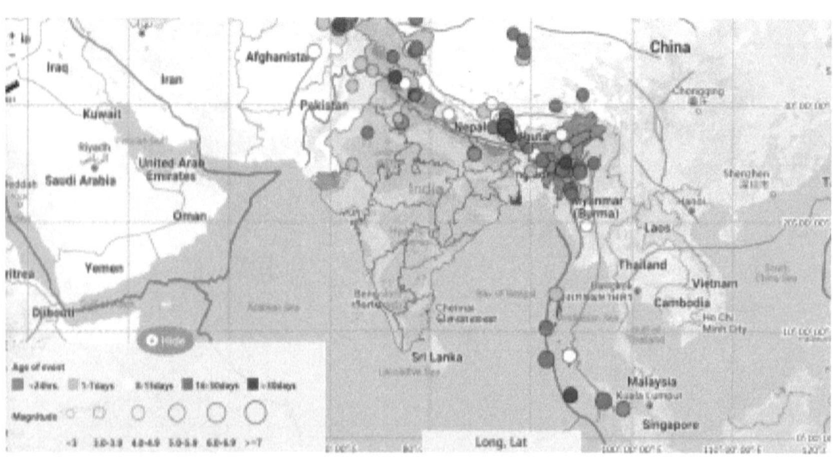

Figure: Earthquakes in India - National Center for Seismology. *https://riseq.seismo.gov.in/riseq/earthquake*

1.2 Why Earthquakes Happen Here

Most of the tremors felt in Delhi originate from the mighty Himalayas, where the Indian tectonic plate is colliding with the Eurasian plate at a speed of nearly 5 cm per year. This collision generates immense pressure along fault lines, which, when released, results in earthquakes. But even beyond the Himalayas, hidden fault lines beneath the Indo-Gangetic Plain occasionally shift, creating unexpected tremors that can shake buildings and rattle nerves.

Figure: Earthquake hazard zone map of India. By Ministry of Earth Science - https://www.moes.gov.in/, GODL-India, https://commons.wikimedia.org/w/index.php?curid=140073082

1.3 The Great Earthquakes of the Region

Delhi has experienced significant earthquakes in recorded history, often originating from the Himalayan belt. Some of these are as follows:

- The 1905 Kangra Earthquake (Magnitude 7.8) devastated large parts of North India.
- The 1950 Assam Earthquake (Magnitude 8.6) was one of the most powerful ever recorded in the region, shaking the entire northeastern part of India.
- The 1991 Uttarkashi Earthquake (Magnitude 6.8) caused significant damage in the Himalayan foothills.
- The 2015 Nepal Earthquake (Magnitude 7.8) was felt strongly in Delhi, sparking concerns about the city's earthquake preparedness.

1.4 Could a Big Earthquake Hit Delhi

Scientists warn that Delhi is overdue for a major earthquake. Studies of historical seismic activity suggest that the buildup of pressure along the Himalayan fault system could trigger a quake of magnitude 7 or higher, which would be devastating for the city's older buildings and dense population. The risk is not just hypothetical. Historical records indicate that earthquakes of this magnitude have struck northern India before, and they will strike again.

1.5 What Can Be Done

While earthquakes cannot be prevented, their impact can be minimized. Some key measures include the following:

- Seismic-resistant construction: Ensuring that new buildings follow earthquake-resistant design standards.
- Retrofitting old structures: Strengthening older buildings to withstand seismic shocks.

- Public awareness: Educating people on emergency preparedness, safe evacuation routes, and what to do during an earthquake.
- Government preparedness: Strengthening early warning systems and response teams.

1.6 Looking Ahead

Delhi's earthquakes are a reminder that we live on a dynamic, ever-changing Earth. The same forces that once lifted the Himalayas and shaped the Indo-Gangetic Plain are still at work today. By understanding the past and preparing for the future, we can ensure that when the earth shakes, we are ready.

Chapter 2: The Himalayas Are Still Growing

The Himalayas are not just old, they are also actively rising. Every year, India moves 5 cm northward, colliding with Asia. This means Everest and Kanchenjunga are still getting taller. In this chapter, we discuss how this affects our rivers, climate, and even earthquakes.

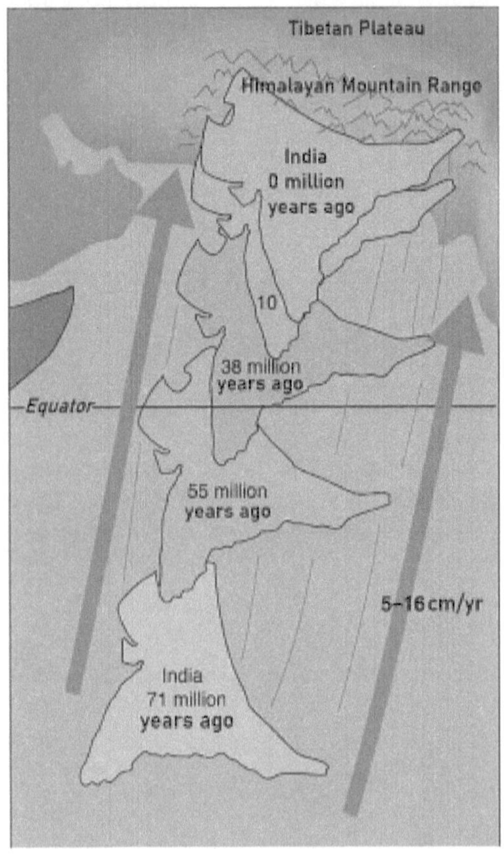

Figure: The movement of the Indian plate toward the Eurasian plate. Haakon Fossen (captions translated into English by Fowler&fowler (talk) 14:27, 15 February 2025 (UTC)), CC BY 3.0 <https://creativecommons.org/licenses/by/3.0>, via Wikimedia Commons

2.1 The Ongoing Rise of the Himalayas

The Himalayas, the tallest and youngest mountain range on Earth, are still growing. This might seem surprising, but the Indian tectonic plate is continuously pushing against the Eurasian plate at a rate of nearly 5 cm per year. This slow but powerful collision causes the mountains to rise, making peaks like Mount Everest and Kanchenjunga taller each year.

2.2 Uttarakhand's Fragile Slopes and Deadly Landslides

Uttarakhand, a picturesque state nestled in the Himalayas, is one of the most landslide-prone regions in India. The reasons for this are as follows:

- **Unstable Young Mountains**: The Himalayas are geologically young, meaning their slopes are not yet well-compacted and are prone to slipping.
- **Monsoon Rains and Erosion**: Heavy rainfall saturates the loose rock and soil, leading to frequent landslides.
- **Deforestation and Construction**: Human activities like road-building, deforestation, and large-scale tourism (especially near Badrinath and Kedarnath) weaken the natural stability of the slopes.

2.3 The Kedarnath Disaster of 2013

One of the most devastating geological tragedies in recent history was the Kedarnath disaster of 2013. A cloudburst triggered massive floods and landslides, leading to thousands of deaths and the destruction of entire villages. Scientists believe that increasing climate instability and unchecked construction in fragile zones contributed to the scale of the disaster.

2.4 How the Himalayas Affect Climate and Rivers

The Himalayas are not just a geological marvel—they shape India's climate. Acting as a barrier to cold winds from Central Asia, they help create the monsoon pattern that defines life in the subcontinent. They also give birth to some of the world's greatest rivers, including the following:

- **Ganges and Yamuna**: Fed by glaciers, these rivers support millions of people.
- **Brahmaputra**: One of the most powerful rivers, prone to massive seasonal floods.
- **Indus**: Historically the cradle of civilization, but now struggling with water-sharing conflicts.

2.5 The Future of the Himalayas

Geologists predict that the Himalayas will continue to rise for millions of years. However, climate change is also accelerating glacial melting, which could lead to increased flooding and instability. As we continue to develop in and around these mountains, balancing growth with geological wisdom will be crucial to prevent further disasters.

Chapter 3: The Giant Indo-Gangetic Plain

Stretching from Punjab to Bengal, the Indo-Gangetic Plain is one of the world's flattest regions. This chapter reveals how millions of years of erosion, glacier melt, and river sediments buried ancient landscapes under layers of soil, creating this vast agricultural heartland.

Figure: Indo Gangetic Plain. Public Domain, https://commons.wikimedia.org/w/index.php?curid=592129

3.1 A Land Shaped by Time

The Indo-Gangetic Plain, stretching from Punjab to Bengal, is one of the world's largest and flattest alluvial landscapes. But how did this vast, almost impossibly even terrain come into existence? The answer lies in millions of years of erosion, glacier melt, and the unrelenting work of rivers shaping the subcontinent.

3.2 The Role of the Himalayas

The Indo-Gangetic Plain owes its existence to the Himalayas, which act as a colossal water tower. Over millions of years, the mountains have been eroded by glaciers, monsoon rains, and strong rivers that carry sediments downstream. The Ganges, Yamuna, Indus, and Brahmaputra rivers have collectively transported massive amounts of rock and silt, burying ancient landscapes under thick layers of sediment. This continuous process has made the land rich for agriculture but has also erased the contours of ancient valleys and mountains that may have once existed.

3.3 The Sediment Factory

The sheer amount of sediment that rivers bring down from the Himalayas is staggering. Every year, the Ganges-Brahmaputra river system alone carries around 1.6 billion tons of sediment to the Bay of Bengal. Over time, this accumulation has built up the flat, fertile plains that support more than 400 million people.

3.4 Why the Indo Gangetic Plain is Flat

Unlike many other river basins, which eventually develop hilly or undulating terrain, the Indo-Gangetic Plain remains astonishingly flat. This is due to the following factors:

- **Constant sediment deposition**: The rivers never stop depositing new layers of silt, smoothing out any variations in the terrain.
- **Tectonic forces**: The region is still geologically active, with subsidence in certain areas keeping the land level.
- **The water table**: The presence of extensive underground aquifers keeps the soil loose and prevents the formation of deep valleys or rock outcrops.

3.5 The Plain as an Agricultural Heartland

This geological gift has turned the Indo-Gangetic Plain into India's agricultural heartland. The region is home to some of the world's most productive farmland, where crops like wheat, rice, sugarcane, and pulses thrive. The alluvial soil, which is soft, nutrient-rich, and easy to cultivate, has supported civilizations for thousands of years, from the Indus Valley to modern-day Punjab and Bihar.

3.6 The Hidden Geological Risks

But this seemingly stable landscape has its share of dangers. Some of them include the following:

- **Flooding**: Because of its low elevation and vast network of rivers, the Indo-Gangetic Plain is highly prone to flooding, especially during the monsoon.
- **Seismic activity**: Though flat, the region sits near major fault lines, making cities like Delhi, Patna, and Lucknow vulnerable to earthquakes.
- **Soil degradation**: Overuse of chemical fertilizers and unsustainable farming techniques threaten the long-term fertility of the plains.

3.7 The Disappearing Rivers and Climate Change

One of the greatest challenges facing the Indo-Gangetic Plain today is climate change. Rising global temperatures are causing Himalayan glaciers to melt at an accelerated rate, which could alter river patterns in unpredictable ways. Some scientists predict that:

- Seasonal flooding will worsen due to heavier glacial melt in the summer.
- Rivers might change course, endangering cities and agricultural

lands.

- In the long run, reduced glacial input could lead to water shortages for millions.

3.8 The Mystery of the Buried Rivers

Geologists and historians have long speculated that beneath the Indo-Gangetic Plain lie the remnants of ancient, lost rivers. One of the most intriguing theories involves the Sarasvati River, mentioned in ancient Hindu scriptures. Satellite imagery has revealed the presence of paleochannels, which are dried-up riverbeds, suggesting that a massive river once flowed through parts of modern-day Haryana and Rajasthan before disappearing, possibly due to tectonic shifts or climate change.

3.9 Past, Present, and Future

This great plain is not just a geological wonder but also the cradle of Indian civilization. Cities have risen and fallen on its banks, empires have thrived on its rich soils, and millions continue to depend on its bounty. However, modern challenges like pollution, groundwater depletion, and climate change threaten its future. Understanding its geology can help us take better care of this essential landscape, ensuring that it remains fertile and habitable for generations to come

Part 2: The Making of India – Supercontinents and Ancient Lands

---

Chapter 4: When India Was Part of Gondwana

Once, India was attached to Madagascar and Antarctica. In this chapter, from fossils in Rajasthan to rocks in Tamil Nadu, we uncover the remnants of India's lost supercontinent past.

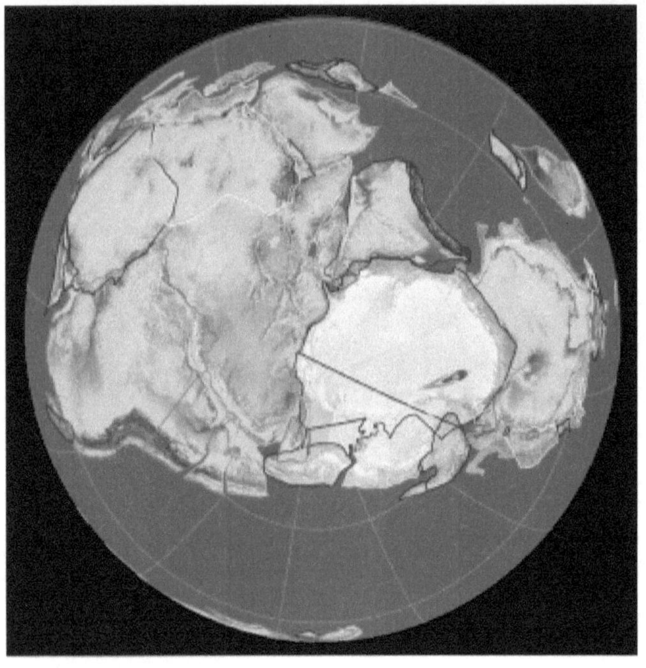

Figure: Gondwana at 420 million years ago, view centered on the South Pole. By Fama Clamosa - Own work, CC BY-SA 4.0, https://commons.wikimedia.org/w/index.php?curid=67001070

Once, India was not a separate landmass but a piece of a much larger puzzle, called Gondwana, the ancient supercontinent. Around 180 million years ago, Gondwana included what are now India, Antarctica,

Africa, South America, Australia, and Madagascar. But as tectonic forces pushed and pulled, this giant landmass broke apart, and India embarked on its dramatic northward journey.

4.1 Clues from the Rocks: Evidence of a Lost World

Today, geologists piece together this prehistoric past by studying the rocks and fossils scattered across India. The Aravalli Range in Rajasthan, one of the oldest mountain systems in the world, contains some of the earliest evidence of Gondwana. Meanwhile, the Cuddapah Basin in Andhra Pradesh holds fossilized remains of primitive life that thrived in this ancient era.

Fossils of Glossopteris, an extinct fern-like plant, found in Rajasthan, Madhya Pradesh, and Odisha, provide a direct link between India and its former Gondwanan neighbors. These fossils are also found in Antarctica, Australia, and Africa, proving that these continents were once connected.

4.2 Madagascar and India: A Lost Connection

If you compare the coastlines of Madagascar and India, they seem to fit together like pieces of a jigsaw puzzle. This is no coincidence—India and Madagascar were joined for millions of years before finally drifting apart around 88 million years ago. Geologists have found matching igneous rock formations and similar dinosaur fossils, such as Majungasaurus, on both landmasses, reinforcing this connection.

4.3 India's Journey Northward: The Birth of the Himalayas

Once separated from Madagascar, India became a solitary traveller, moving at a speed of 15 cm per year, which was one of the fastest recorded tectonic movements in Earth's history. Around 50 million years ago, it collided with the massive Eurasian plate, crumpling and pushing up the land to form the Himalayas, the world's highest mountain range.

4.4 Gondwana's Influence on India's Resources

The ancient Gondwana landscape didn't just shape India's geography. It also left behind vast deposits of valuable resources, including the following:

- **Coal in Jharkhand, Odisha, and Chhattisgarh**: Formed from the lush forests that thrived in the swampy Gondwanan environment.
- **Gold in Karnataka**: Deposited over millions of years in the remnants of ancient river systems.
- **Diamonds in Madhya Pradesh**: Created under immense pressure deep within the Earth's crust.

4.5 The Legacy of Gondwana in Modern India

Gondwana may have broken apart millions of years ago, but its impact on India is still visible today. The Western Ghats, stretching along the west coast, owe their origin to this ancient supercontinent. The Satpura and Vindhya ranges, which divide northern and southern India, were shaped by tectonic forces dating back to Gondwana's breakup.

Even India's rich wildlife carries echoes of this prehistoric past. Many species found in the Western Ghats and Eastern Himalayas have close relatives in Africa and Australia, highlighting their shared Gondwanan ancestry.

4.6 What If Gondwana Had Not Broken Apart

Imagine an alternate reality where Gondwana had remained intact. India would still be connected to Antarctica, possibly covered in glaciers. There would be no Himalayas, no monsoon-driven rivers like the Ganges, and an entirely different climate and ecosystem.

The breakup of Gondwana was not just a geological event, rather it set the stage for India's unique landscapes, climate, and biodiversity, shaping the subcontinent we know today.

Chapter 5: Sri Lanka, Kerala's Lost Twin

Sri Lanka isn't just close to India. It is geologically related. The rocks in Sri Lanka and Kerala are almost identical, hinting at a time when they were one landmass. In this chapter, we learn why they separated, and what this tells us about Earth's shifting plates.

5.1 A Shared Geological History

Sri Lanka and Kerala appear almost like separated twins on a map, and their geological past confirms this. Both landmasses were once connected as part of the ancient Gondwana supercontinent. When India drifted away from Madagascar and Antarctica, Sri Lanka remained attached for a while before eventually separating.

5.2 The Western Ghats and Sri Lanka's Highlands

Geologically, Sri Lanka's highlands are an extension of the Western Ghats. The Precambrian rocks found in Sri Lanka's central hills are nearly identical to those of Kerala. This shared history explains similarities in terrain, flora, and fauna between the two regions.

5.3 Why Didn't Sri Lanka Move Away Like Madagascar

Unlike Madagascar, which moved farther away, Sri Lanka remained close to the Indian subcontinent due to lower tectonic activity in the region. The shallow Palk Strait, which separates India and Sri Lanka, was once a land bridge, allowing species migration and human movement.

5.4 Geological and Ecological Challenges Today

Such challenges include the following:

- **Rising Sea Levels**: Climate change threatens the low-lying coastal areas of both Kerala and Sri Lanka.
- **Sand Mining**: Unregulated mining in Kerala and Sri Lanka has led to erosion and ecosystem damage.
- **Deforestation**: The biodiversity-rich Western Ghats and Sri Lanka's rainforests are under constant pressure from human activities.

5.5 Moving Forward

Understanding the geological connection between Kerala and Sri Lanka helps us appreciate our shared natural heritage. Sustainable conservation policies can help preserve the landscapes and biodiversity of these unique regions for future generations.

Part 3: The Fiery Past – Volcanoes and the Deccan Mystery

Chapter 6: The Deccan Traps

Before dinosaurs disappeared, western India was drowning in lava. The Deccan Traps eruption was one of Earth's biggest volcanic events. In this chapter, we discuss how it shaped today's landscapes, and whether it played a role in wiping out the dinosaurs.

Figure: Oblique satellite view of the Deccan Traps. Planet Labs, Inc, CC BY-SA 4.0 <https://creativecommons.org/licenses/by-sa/4.0>, via Wikimedia Commons

6.1 The Birth of the Deccan Traps

Nearly 66 million years ago, a series of massive volcanic eruptions shook the Indian subcontinent. These eruptions, spread over thousands of years, spewed vast amounts of lava across what is now western and central India, forming the Deccan Traps—one of the largest volcanic features on Earth. Today, the region covers over 500,000 square kilometres, forming the rugged Deccan Plateau.

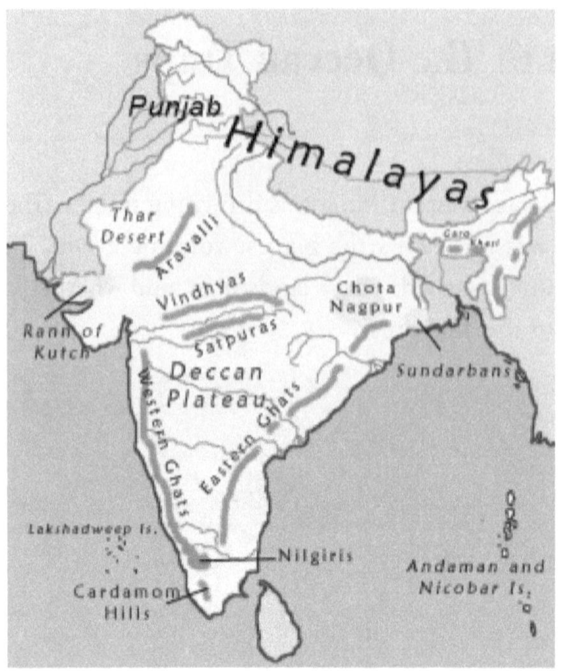

Figure: Regions of India, showing the Deccan Plateau surrounded by the various mountain ranges. Nichalp, CC BY-SA 3.0 <http://creativecommons.org/licenses/by-sa/3.0/>, via Wikimedia Commons

6.2 A Land Forged by Fire

The Deccan Plateau, a vast expanse covering most of central and southern India, owes its existence to one of the most colossal volcanic eruptions in Earth's history. Around 66 million years ago, during the end of the Cretaceous period, relentless lava flows spewed from the Earth's crust, blanketing the land in layer upon layer of basaltic rock. Over millions of years, weathering and erosion transformed this hardened volcanic rock into the fertile yet distinctively red and black soils that define the region today.

6.3 The Deccan and the Dinosaur Extinction

A fascinating theory connects the Deccan Traps eruptions to the mass extinction event that wiped out the dinosaurs. Some scientists believe that the massive release of carbon dioxide and sulphur dioxide from these eruptions caused dramatic climate shifts, leading to environmental disasters that contributed to the dinosaurs' demise.

6.4 Looking Ahead: Preserving the Deccan

To ensure the sustainability of this geologically significant region, conservation efforts are needed. Some of them are as follows:

- Promoting soil conservation techniques to combat erosion.
- Implementing better water management practices.
- Enforcing environmental regulations on mining and industrial activities.

The Deccan Plateau is not just a remnant of Earth's fiery past. It continues to shape the lives and livelihoods of millions. Understanding its history and protecting its resources is key to a sustainable future.

Chapter 7: What Lies Beneath the Deccan Plateau

The Deccan's soil is famously red and black. What makes it that way? In this chapter, we explore how ancient lava flows created India's rich farmlands and the mineral magic behind their colors.

7.1 Why Is the Soil Red in Deccan

The striking red soil of the Deccan is the result of iron-rich basaltic rock oxidizing over time. The oxidation of iron, much like rusting metal, imparts a reddish hue to the soil. This process occurs in hot and humid climates, which are common in much of the Deccan. However, red soil is not as fertile as black soil because it lacks organic matter and nitrogen, requiring careful cultivation techniques to support agriculture.

7.2 The Fertile Black Cotton Soil

In contrast, the black soil, also known as Regur soil, retains moisture exceptionally well, making it ideal for crops like cotton, millets, and pulses. This soil originated from volcanic ash deposits and is rich in minerals such as calcium, magnesium, and potash. Unlike red soil, black soil has a high clay content, which allows it to hold water longer, benefiting crops during dry seasons.

7.3 Ancient Rivers and Soil Formation

Millions of years of rain, wind, and river activity shaped the Deccan Plateau's soil. The Godavari, Krishna, and Narmada rivers carried sediments across the plateau, distributing minerals and nutrients over vast areas. This created a patchwork of red, black, and lateritic soils, each with distinct properties that influence modern agriculture and water retention.

Figure: Red soil in India. By Carla Antonini - Own work, CC BY-SA 3.0, https://commons.wikimedia.org/w/index.php?curid=20164061

7.4 The Secrets Behind the Red Soil

One of the most striking features of the Deccan Plateau is its vibrant red and black soil.

- **Red Soil**: The presence of iron oxide gives Deccan soil its characteristic reddish hue. This type of soil is common in semi-arid regions and is less fertile due to lower organic content.
- **Black Soil (Regur Soil)**: Formed from weathered basalt, this soil is rich in minerals and highly moisture-retentive, making it ideal for crops like cotton.

7.5 The Impact on Agriculture

The Deccan's unique soils have played a crucial role in India's agricultural development. Black soil, known for its ability to hold water, supports cotton farming, while red soil is used for cultivating millets, pulses, and oilseeds. However, climate change and overuse of land have led to increasing soil degradation, requiring sustainable farming practices to preserve its fertility.

7.6 The Role of the Monsoon

The monsoon plays a crucial role in shaping Deccan soils. Seasonal rainfall triggers erosion and nutrient redistribution, making some regions fertile while depleting others. The combination of volcanic history and monsoon-driven soil dynamics makes the Deccan Plateau one of the most agriculturally complex regions in the world.

7.7 Challenges for Modern Agriculture

While the Deccan's soils have supported farming for thousands of years, modern agricultural practices pose serious challenges:

- **Soil Erosion**: Deforestation and overgrazing have led to severe soil degradation, reducing crop yields.
- **Water Scarcity**: The plateau's high water retention is a double-edged sword; black soil cracks when dry, making irrigation critical.
- **Nutrient Depletion**: Excessive use of chemical fertilizers has stripped the soil of its natural fertility, leading to long-term sustainability concerns.

7.8 Can the Deccan's Soil Support the Future

Scientists and environmentalists are working on sustainable farming techniques to preserve soil fertility, some of which include the following:

- **Organic farming** is being promoted to restore microbial activity in depleted soils.
- **Rainwater harvesting** techniques are helping farmers combat

water scarcity.

- **Crop rotation and natural fertilizers** are being used to enrich the land without depleting nutrients.

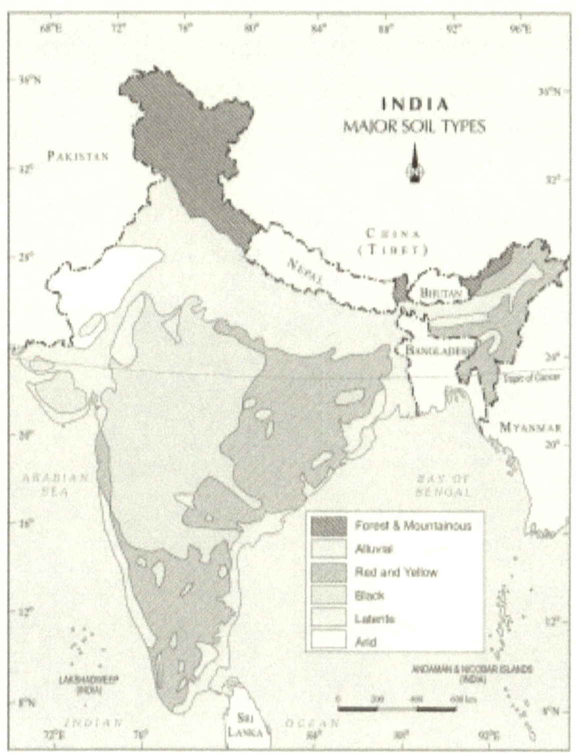

Figure: Major soil types in India. ncert, CC BY-SA 3.0 <https://creativecommons.org/licenses/by-sa/3.0>, via Wikimedia Commons

7.9 The Deccan's Red and Black Soil in Industry and History

The rich mineral composition of the Deccan Plateau has not only supported agriculture but also fuelled India's industrial growth. The region is abundant in iron, manganese, and bauxite, making it an important hub for mining. Additionally, the Ajanta and Ellora caves, carved out of Deccan basalt, showcase how geology has influenced India's architectural and artistic history.

7.10 A Geological Wonder Beneath Our Feet

The Deccan Plateau's soils are more than just dirt. They are a living record of Earth's fiery past, a key to India's agricultural economy, and a resource that must be managed carefully for future generations. Understanding the origins and importance of red and black soil gives us a deeper appreciation for the land beneath our feet, reminding us that even the simplest things, like the colour of the earth, hold stories millions of years in the making.

Part 4: Coasts, Islands, and Forgotten Lands

Chapter 8: India's Ever-Changing Coastlines

From the shifting sands of Gujarat to the sinking lands of the Sundarbans, India's coasts are in flux. In this chapter, we discuss how did they form, and how will they change in the future with rising sea levels.

Figure: Sea erosion at Valiyathura Kerala, India. BHAVAPRIYA J U, CC BY-SA 4.0 <https://creativecommons.org/licenses/by-sa/4.0>, via Wikimedia Commons

8.1 The Ever-Changing Shoreline

India's extensive coastline, stretching over 7,500 kilometres, is one of the most dynamic landscapes in the country. From the sandy shores of Goa to the mangrove swamps of the Sundarbans, the coast is constantly shaped by ocean currents, tides, and monsoons. But today, coastal erosion, unchecked development, and climate change pose significant challenges to these regions.

8.2 Coastal Erosion: A Growing Concern

Modern India is also facing the threat of disappearing land due to rising sea levels and climate change. Regions like Sundarbans in West Bengal, Chilika Lake in Odisha, and the coastal villages of Kerala are witnessing alarming rates of land erosion and submersion. If global temperatures continue to rise, entire communities may be displaced, echoing the lost lands of the past.

Some of India's most famous beaches and shorelines are shrinking due to natural and human-induced erosion. Factors contributing to this include the following:

- **Rising sea levels** due to climate change.
- **Sand mining** disrupting natural sediment replenishment.
- **Construction of ports and sea walls**, altering natural coastal dynamics.

States like Tamil Nadu, Kerala, and Odisha are witnessing severe coastline loss, impacting local communities, agriculture, and fisheries.

8.3 The Impact of Rising Seas On Coastal Cities

With Mumbai, Chennai, and Kolkata sitting at low elevations, rising sea levels are a significant concern. Predictions suggest that by 2100, many parts of these cities could be underwater, displacing millions. Flooding, saltwater intrusion into freshwater sources, and loss of biodiversity are major threats.

8.4 The Sundarbans: A Mangrove Ecosystem at Risk

The Sundarbans, home to the Bengal tiger and a rich mangrove ecosystem, is one of the most vulnerable regions to sea-level rise. The delicate balance of saline and freshwater is being disrupted, affecting wildlife and local fishing communities. Deforestation and human encroachment further exacerbate the situation.

8.5 Finding a Balance: Sustainable Coastal Development

While development is necessary for economic growth, balancing it with sustainability is critical. Steps that can help include the following:

- **Coastal Regulation Zones (CRZs)** to prevent harmful construction.
- **Restoration of mangroves and wetlands** to act as natural buffers.
- **Use of renewable energy sources**, like offshore wind and tidal power, to reduce environmental impact.

8.6 How to Preserve India's Coastal Heritage

Efforts are underway to protect India's coastal and submerged heritage through the following ways:

- **Underwater Archaeology**: Exploring and documenting submerged ruins before they are lost forever.
- **Coastal Conservation**: Planting mangroves, reducing sand mining, and enforcing stricter coastal protection laws.
- **Climate Adaptation**: Developing strategies to protect coastal communities from rising waters.

India's coastlines are its lifeline, supporting millions of livelihoods. Protecting them from ecological degradation and climate risks is crucial for the country's future.

Chapter 9: Andaman & Nicobar

The Andaman and Nicobar Islands are living geology. This chapter explores the volcanic origins of these islands and how the 2004 tsunami revealed the power of shifting plates beneath the sea.

Figure: Barren Island (a part of Andamans) erupts. Arijayprasad, CC BY-SA 4.0 <https://creativecommons.org/licenses/by-sa/4.0>, via Wikimedia Commons

9.1 The Volcanic Birth of the Islands

The Andaman and Nicobar Islands are not just picturesque tropical paradises; they are a geological hotspot. Formed by the subduction of the Indo-Australian plate beneath the Eurasian plate, these islands are part of an active tectonic region known as the Sunda Subduction Zone. The volcanic origins of these islands have resulted in unique landscapes and rich biodiversity.

9.2 Earthquakes, Tsunamis, and a Shifting Landscape

Being located near a tectonic boundary, the Andaman and Nicobar Islands are highly susceptible to earthquakes and tsunamis. The 2004 Indian Ocean tsunami, triggered by a massive undersea earthquake, caused immense destruction in this region, submerging some land while uplifting others. The geological movements in this area continue to alter its topography.

9.3 The Changing Climate and Rising Seas

Contemporary concerns in the Andaman and Nicobar Islands include rising sea levels due to climate change. Many low-lying areas are at risk of being submerged, forcing local communities to adapt or relocate. Coastal erosion and habitat loss threaten both human settlements and wildlife.

9.4 Balancing Development and Conservation

The islands' strategic location makes them critical for national security and economic development, yet unchecked construction and tourism pose risks to their fragile ecosystems. Sustainable planning, improved disaster preparedness, and ecological conservation efforts are necessary to preserve the unique geological and ecological character of the Andaman and Nicobar Islands.

Chapter 10: Lost Lands of India

From the ancient land of Kumari Kandam to submerged temple ruins, this chapter uncovers India's lost landscapes.

10.1 Submerged Civilizations and Vanished Landscapes

India's landmass has changed dramatically over the millennia, shaped by tectonic shifts, rising sea levels, and climatic changes. Entire cities and regions now lie underwater: from the legendary Dwarka, believed to have been swallowed by the sea, to the fabled Kumari Kandam, thought to be a lost landmass in the Indian Ocean, India has its fair share of submerged secrets.

Figure: A painting of ancient Dwarka city, now said to be submerged underwater. Grindlay's, Public domain, via Wikimedia Commons

10.2 The Mystery of Dwarka: Fact or Fiction

Ancient texts describe Dwarka, the city of Lord Krishna, as a magnificent urban centre that eventually sank beneath the sea. Archaeological excavations off the coast of Gujarat have revealed stone structures, artifacts, and pottery, suggesting the presence of an advanced settlement thousands of years ago. Some researchers believe that changing coastlines and a powerful tsunami may have contributed to its submersion, adding a geological dimension to this mythological tale.

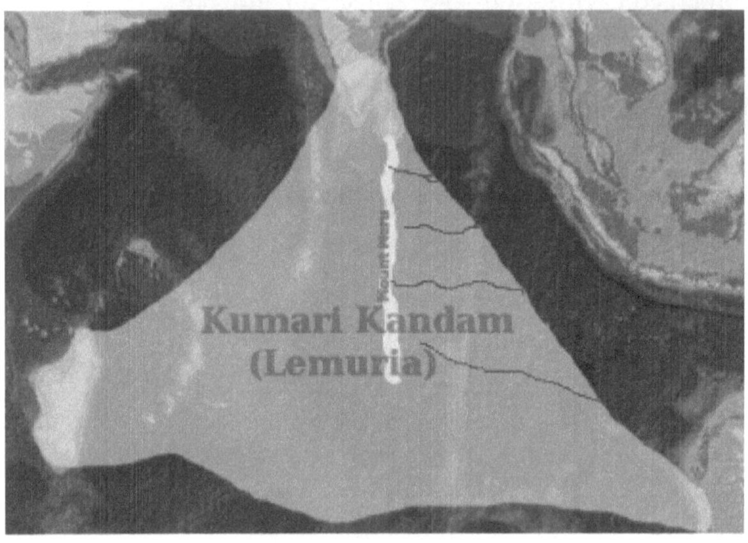

Figure: Kumari Kandam (Lemuria) Map. CC BY-SA 3.0, https://commons.wikimedia.org/w/index.php?curid=1889920

10.3 Kumari Kandam: India's Atlantis

According to Tamil legends, Kumari Kandam was a vast landmass south of India, home to an ancient Tamil civilization. While there is no concrete geological evidence for this lost continent, studies suggest that the sea levels were much lower thousands of years ago, exposing land that is now submerged. The submerged remnants off the southern coast of India provide intriguing clues to what may have once existed.

10.4 The Sunken Temples of Mahabalipuram

In 2004, after the devastating Indian Ocean tsunami, eyewitnesses reported seeing stone structures emerge from the sea off the coast of Tamil Nadu. Subsequent underwater explorations revealed evidence of ancient temples, confirming long-standing local legends of a city lost to the ocean. These discoveries highlight how natural disasters and geological processes can dramatically alter coastlines and bury human history beneath the waves.

10.5 The Science Behind Disappearing Lands

Several geological factors contribute to the submergence of land. Some of them are as follows:

- **Tectonic Activity**: The movement of Earth's crust can lead to sudden shifts, causing parts of land to sink below sea level.
- **Rising Sea Levels**: Climate change and melting glaciers have raised ocean levels, submerging low-lying regions over time.
- **Erosion and Coastal Subsidence**: Constant wave action, river flooding, and monsoons erode shorelines, leading to the gradual disappearance of land.

The lost lands of India are more than just ancient legends—they are a reminder of our planet's ever-changing landscape and the power of natural forces. While we may never fully recover what has been submerged, understanding these lost worlds helps us appreciate the forces shaping our present and future.

Part 5: How It Affects Us – The Human Connection

Chapter 11: Earthquakes and Everyday Life

How can understanding geology help us prepare for future earthquakes? This chapter explains seismic zones, why some cities are more vulnerable, and how to build for the future.

11.1 Living on a Shaking Land

Earthquakes have shaped India's history, geography, and cities in ways few people realize. From the towering Himalayas to the crowded streets of Delhi, seismic activity is an invisible force that threatens millions of lives. But understanding why earthquakes happen, which cities are most vulnerable, and how we can prepare can mean the difference between disaster and survival.

11.2 Why Does India Experience So Many Earthquakes

India sits on a geological collision zone—the boundary where the Indian Plate crashes into the Eurasian Plate. This ongoing tectonic movement is responsible for:

- The **rise of the Himalayas** (which are still growing!).
- The **frequent earthquakes in northern India, Nepal, and Pakistan.**
- The **occasional tremors in central and southern India** due to ancient fault lines.

Scientists classify India into four seismic zones, from Zone II (least risk) to Zone V (highest risk). The Himalayan region, Northeast India, and parts of Gujarat fall into Zone V, making them especially prone to devastating earthquakes.

11.3 The Cities Most at Risk

Some Indian cities are more vulnerable to earthquakes than others due to their location, population density, and building structures:

- **Delhi** – Lies near major fault lines and experiences tremors frequently.
- **Srinagar** – Nestled in the Himalayas, one of the most active seismic regions.
- **Guwahati** – In the Northeast, which sits on multiple faults and is highly prone to quakes.
- **Shimla & Dehradun** – Located in the fragile Himalayan belt.
- **Bhuj (Gujarat)** – The 2001 earthquake killed 20,000 people and reshaped the region.

Many of these cities have poorly constructed buildings, increasing the risk of damage and casualties during a major quake.

11.4 How Can We Build for a Safer Future

Modern earthquake engineering offers solutions to reduce the destruction caused by earthquakes:

- **Flexible Buildings**: Japan and California use structures that sway instead of collapsing.
- **Base Isolation Technology**: Buildings are placed on shock-absorbing foundations that absorb seismic energy.
- **Retrofitting Old Buildings**: Strengthening existing structures with modern materials makes them more resilient.
- **Zoning Laws**: Governments must enforce building codes, ensuring new constructions can withstand quakes.

India is slowly adopting these measures, but many older structures remain dangerously unprepared.

11.5 The Devastating Earthquakes of India: Lessons from the Past

Some of India's worst earthquakes remind us why preparation is critical:

- **2001 Bhuj Earthquake (Gujarat, 7.7 magnitude)** – Flattened towns, killed 20,000, and left 600,000 homeless.
- **2015 Nepal Earthquake (felt across North India, 7.8 magnitude)** – Destroyed Kathmandu, shook Delhi, and killed nearly 9,000 people.
- **1934 Bihar-Nepal Earthquake (8.0 magnitude)** – One of the deadliest in Indian history, razing Patna and parts of Nepal.
- **1993 Latur Earthquake (6.2 magnitude, Maharashtra)** – A rare event in peninsular India that claimed nearly 10,000 lives.

Each of these disasters exposed weak buildings, slow response times, and the urgent need for better earthquake preparedness.

11.6 Can We Predict Earthquakes

Unlike hurricanes or floods, earthquakes strike without warning. Scientists are working on prediction models using:

- Seismic patterns – Studying past earthquakes to find trends.
- Animal behavior – Some animals sense tremors before they happen.
- GPS Monitoring – Tracking tiny shifts in tectonic plates.

While no perfect prediction system exists, early warning networks can detect tremors seconds before they intensify, giving people time to seek shelter.

11.7 What Should You Do in an Earthquake

If an earthquake strikes, your response can save your life:

- Inside a building: Drop, Cover, and Hold On. Take shelter under a sturdy table.
- Outside: Stay away from trees, buildings, and power lines.
- Driving: Stop safely and avoid bridges or overpasses.
- Near the coast: Move to higher ground immediately to avoid tsunamis.

Preparedness also means having:

- A disaster kit with food, water, and medicine.
- A family emergency plan.
- Knowledge of safe spots in your home or workplace.

11.8 The Future of Earthquake Safety in India

With rapid urbanization, India must act now to make its cities earthquake-resilient:

- **Stricter Building Codes** – Ensuring new structures can withstand strong tremors.
- **Public Awareness Campaigns** – Teaching communities how to prepare.
- **Early Warning Systems** – Implementing seismic detection networks like Japan and the US.
- **Emergency Response Training** – Training citizens and officials in rescue operations.

By combining scientific research, engineering innovations, and public education, India can reduce earthquake damage and save lives.

11.9 Conclusion: Living with Earthquakes

Earthquakes are an inevitable part of India's geology, but disaster does not have to be. With the right planning, technology, and awareness, we can build a future where tremors no longer mean tragedy.

Chapter 12: The Wealth Beneath Our Feet

India's geological riches—gold in Karnataka, coal in Jharkhand, diamonds in Panna—have shaped economies and history. In this chapter, we discuss the future of resource mining.

INDIA

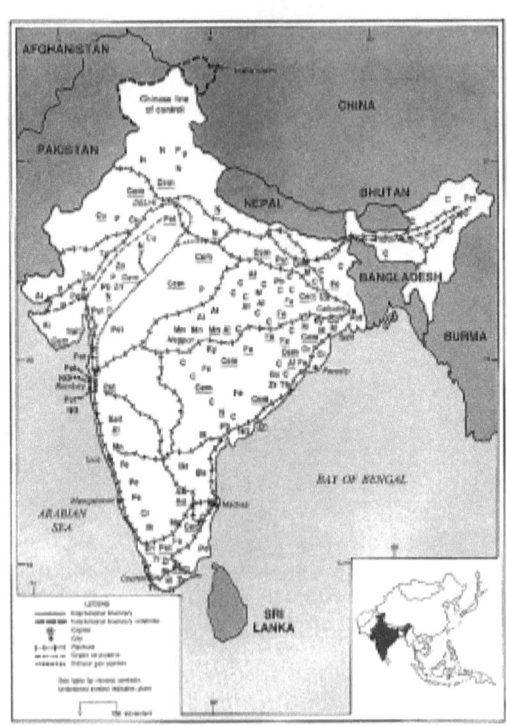

Figure: U.S. Geological Survey Map of the mineral deposits in India. United States Geological Survey via Wikimedia commons.

12.1 A Land Rich in Minerals

Beneath India's diverse landscapes lies an incredible geological treasure trove—gold in Karnataka, coal in Jharkhand, diamonds in Madhya Pradesh, and iron ore in Odisha. These resources have shaped India's economy, fueled industries, and even played a role in shaping historical conflicts and global trade routes.

Figure: Champion Reef mine shaft at Kolar Gold Fields. Adu65 at English Wikipedia, Public domain, via Wikimedia Commons

12.2 Gold: Karnataka's Ancient Treasure

Karnataka's Kolar Gold Fields (KGF) and Hutti Gold Mines are among the oldest and richest gold-producing regions in India. The gold deposits here date back nearly 2.5 billion years, formed deep within the Earth's crust due to intense geothermal activity. Although KGF has been largely exhausted, India still has significant untapped gold reserves waiting to be explored.

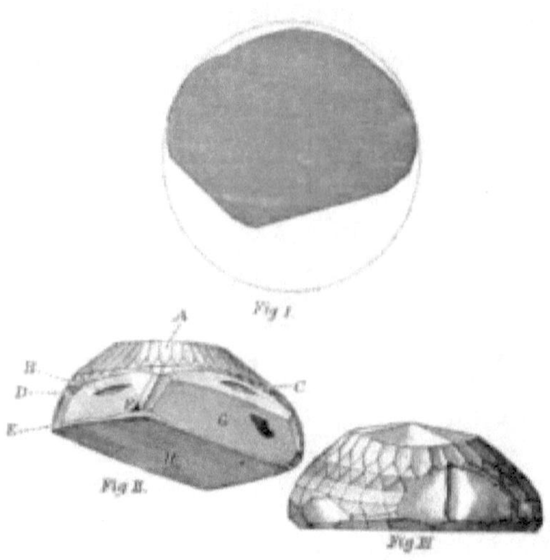

Figure: Diagram of the pre-1852 cut Koh-i-Noor. Jean-Baptiste Tavernier, Public domain, via Wikimedia Commons

12.3 Diamonds in Panna: India's Gem of Kings

India was once the world's largest diamond producer, with the Golconda mines in Telangana producing legendary gems like the Koh-i-Noor. Today, Panna in Madhya Pradesh remains one of the few places where diamonds are still mined. These diamonds are formed over millions of years under extreme heat and pressure deep beneath the Earth's surface.

12.4 The Black Gold: India's Coal Reserves

Coal is the backbone of India's energy sector, with Jharkhand, Odisha, and Chhattisgarh housing some of the richest coal deposits. These reserves were formed 300 million years ago, during the Carboniferous period, when dense forests were buried under sediments and gradually transformed into coal. However, with growing concerns about pollution and climate change, India is now balancing coal mining with a shift towards renewable energy.

12.5 Iron Ore: The Foundation of India's Industry

Iron ore, crucial for steel production, is found in abundance in Odisha, Chhattisgarh, and Karnataka. The Bailadila Hills in Chhattisgarh and the Barbil-Keonjhar belt in Odisha supply iron ore for both domestic industries and exports, making India a major player in the global steel market.

12.6 The Environmental Cost of Mining

While India's mineral wealth has fueled economic growth, it has also led to deforestation, land degradation, and displacement of local communities. The impact of mining includes:

- **Water pollution**: Runoff from mines contaminates rivers and groundwater.
- **Loss of biodiversity**: Mining disrupts natural ecosystems, endangering wildlife.
- **Human displacement**: Indigenous communities often lose their ancestral lands to large-scale mining operations.

12.7 The Future of Resource Extraction in India

With depleting reserves and rising environmental concerns, the future of mining in India will depend on sustainable practices such as:

- **Green mining technology**: Using eco-friendly methods to extract minerals with minimal environmental damage.
- **Recycling metals**: Reducing dependency on mining by increasing metal recycling.
- **Exploring deep-sea mining**: India's coastline holds untapped mineral resources on the ocean floor, offering new possibilities if done responsibly.

12.8 Striking a Balance: Resource Extraction vs. Conservation

India's mineral wealth is a double-edged sword—essential for economic growth yet fraught with environmental challenges. As we move toward a future of renewable energy and sustainability, the mining industry must adapt to ensure that the wealth beneath our feet benefits both present and future generations.

Chapter 13: Living with Geology

Geology is not just about the past; it affects our future too. This chapter explores how India's landscapes shape agriculture, water supply, and even tourism.

Figure: NASA satellite photo of Adam's Bridge or Ram Setu, connecting India and Sri Lanka: India on top, Sri Lanka at the bottom. National Aeronautics & Space Administration, Public domain, via Wikimedia Commons

13.1 Adam's Bridge: Myth and Geology

Adam's Bridge, or Ram Setu, is a chain of limestone shoals connecting India and Sri Lanka. Geological evidence suggests that this formation was once a land bridge, possibly walkable during lower sea levels. While its origins remain debated, satellite imagery and geological studies indicate that shifting sea levels and sediment deposits have played a role in its current form. The bridge's significance extends beyond science, forming an intersection of history, mythology, and natural forces.

13.2 How Rocks Shaped Indian Culture

From ancient temples carved out of rock to the strategic positioning of cities, geology has influenced the course of Indian civilization. The availability of granite, sandstone, and limestone dictated the construction of magnificent structures like the temples of Mahabalipuram, the forts of Rajasthan, and the caves of Ajanta and Ellora.

Geology is not just about the past. It continues to shape our present and future. By understanding it, we can build a more sustainable and resilient India.

13.3 Geology and India's Water Crisis

The geology of a region directly determines how much water is available to its people. In India, the type of rock beneath our feet determines whether a village has access to groundwater or suffers from drought. Fractured granite rocks in southern India hold limited groundwater, while the porous alluvial sediments of the Indo-Gangetic Plain act as vast underground reservoirs. Understanding these geological differences is essential for sustainable water management.

India is facing a serious groundwater crisis. Over-extraction of water from underground aquifers for agriculture and industry has caused water tables to drop alarmingly in states like Punjab, Haryana, and Rajasthan. In many regions, farmers now dig wells hundreds of metres deep to reach water that took thousands of years to accumulate. Once these aquifers are depleted, they may take centuries to recharge. Geological surveys play a vital role in identifying sustainable water sources and guiding conservation efforts.

13.4 Geology and Tourism: India's Geological Wonders as Destinations

India's geological heritage is also a source of significant tourism and scientific interest. Geotourism, travel that focuses on geological features and landscapes, is a growing field that can bring economic benefits to local communities while raising awareness about Earth's history. Some remarkable geological tourism destinations in India include the following:

- **Lonar Crater, Maharashtra:** One of the few impact craters in the world formed in basaltic rock, created by a meteorite strike roughly 50,000 years ago. It now contains a saline lake and is a UNESCO-recognised geological heritage site.
- **Borra Caves, Andhra Pradesh:** Ancient limestone caves in the Araku Valley, featuring spectacular stalactites and stalagmites formed over millions of years by underground water slowly dissolving the rock.
- **Fossil Park, Dholavira, Gujarat:** An extraordinary site near the ruins of the ancient Indus Valley city of Dholavira, where fossil wood and marine fossils bear witness to the dramatic changes in this landscape over millions of years.
- **Spiti Valley, Himachal Pradesh:** A cold desert valley at high altitude containing some of the best-preserved marine fossils in the world, including ammonites and brachiopods that lived in the ancient Tethys Sea long before the Himalayas rose from its bed.

13.5 Geology and Agriculture: The Soil Beneath the Fields

India's geological diversity is directly reflected in its agricultural diversity. Every soil type found across the country has a geological origin that determines what crops will grow and what farming methods are best suited to the land. The deep black cotton soils of Maharashtra and Madhya Pradesh, born from ancient basalt, are ideal for cotton and sorghum. The light red soils of the Eastern Ghats support cashews and groundnuts. The rich alluvial plains of the Ganga basin feed hundreds of millions through wheat and rice cultivation.

Understanding the geological origins of these soils helps farmers make better decisions about irrigation, fertilisation, and crop rotation. It also guides government policy on land use, helping to prevent the kind of soil degradation that can turn productive farmland into wasteland within a generation. In this sense, geology is not merely an academic subject: it is the foundation of food security for over a billion people.

Chapter 14: The Future of India's Land

Will India's landscapes change in the coming centuries? How will climate change affect our mountains, plains, and coastlines? In this chapter, we look ahead at the challenges and opportunities.

14.1 The Changing Landscape: What Lies Ahead

India's terrain is constantly evolving. While we often think of landscapes as permanent, geology tells us otherwise. Over the coming centuries, India's mountains, plains, and coastlines will undergo profound transformations due to climate change, tectonic activity, and human impact.

14.2 The Himalayas: Still Rising, Still Shaking

The Himalayas are still growing, pushed higher each year by the relentless collision of the Indian and Eurasian plates. The big question remains, whether we can balance development and conservation in this fragile region.

14.3 The Indo-Gangetic Plain: A Future of Floods and Water Crises

The Indo-Gangetic Plain, home to over 400 million people, is also vulnerable to environmental shifts. The challenge is how to manage our water resources sustainably, in a future where water scarcity could become the norm.

14.4 The Deccan Plateau: A Land Drying Up

The Deccan Plateau, shaped by ancient volcanic activity, is facing desertification in some areas due to deforestation, erratic monsoons and soil exhaustion. The opportunity lies in using reforestation and regenerative agriculture to restore the Deccan's soil and water systems.

14.5 India's Sinking Coastlines and Vanishing Islands

The rising Indian Ocean is claiming land along India's coastline. Coastal cities like Mumbai, Chennai, and Kolkata face increased flooding, while the Sundarbans, home to the Royal Bengal Tiger, are slowly drowning. Coastal protection measures, mangrove restoration, and climate adaptation strategies are needed to save these vulnerable regions.

14.6 The Hidden Future: What Lies Beneath

As land disappears, new landscapes emerge. Future explorers may uncover lost rivers, buried civilizations, and new mineral deposits beneath our shifting land. The study of geoarchaeology, combining geology and archaeology, could reveal incredible secrets about India's past while helping us prepare for its future.

14.7 The Path Forward: Can India Adapt

The forces shaping India's land are beyond human control, but our response to them is not. Sustainable policies, scientific research, and community action can help mitigate the worst impacts. Some key solutions include the following:

- **Disaster preparedness**: Early warning systems for earthquakes, floods, and cyclones can save lives.
- **Sustainable land use**: Smarter urban planning can prevent groundwater depletion and land erosion.
- **Conservation efforts**: Reforestation, wetland protection, and eco-friendly farming can help preserve our landscapes.

14.8 India in 500 Years: A Land Transformed

Looking centuries ahead, India's geography may look very different. The Himalayas will be taller, parts of the Indo-Gangetic Plain may be underwater, and new river systems may emerge. But one thing is certain: the land will continue to change, just as it has for millions of years. Understanding these shifts can help us prepare for the challenges and opportunities ahead.

The future of India's land is not just about geology. It is about human resilience, adaptation, and the choices we make today. By respecting the natural forces that shape our world, we can ensure that India's landscapes remain vibrant and life-sustaining for generations to come

Chapter 15: Conclusion

India's landscape is a living storybook, one written over billions of years in rock, water, and fire. From the towering Himalayas to the flat Indo-Gangetic Plain, from the volcanic Deccan Plateau to the sinking coastal lands, the country's geology is in constant motion—shaped by tectonic forces, climatic shifts, and human activity.

15.1 Summary of Learnings

Throughout this book, we have uncovered the deep geological forces that built India, including the following:

- **Tectonic collisions** that birthed the Himalayas and continue to cause earthquakes.
- **Erosion and sedimentation** that formed the vast Indo-Gangetic Plain, the cradle of Indian civilization.
- **Ancient volcanic eruptions** that shaped the Deccan Plateau, leaving behind fertile yet fragile soil.
- **Submerged lands and shifting coastlines**, where rising sea levels and climate change threaten ancient cities and modern settlements.
- **India's rich mineral wealth**, which has fuelled industries but also raised concerns over sustainability.

Yet, despite these forces, India's landscape is not just a relic of the past—it is a dynamic, evolving terrain that continues to shape human life.

15.2 The Challenges Ahead

The geological forces that built India are still at work, but today they interact with a new player, which is human activity. Climate change, deforestation, over-extraction of resources, and urban expansion are rapidly altering the landscape in ways that were unimaginable a few centuries ago.

Key issues that need urgent attention include the following:

- **Earthquake preparedness** in high-risk zones like Delhi, Uttarakhand, and the Northeast.
- **Sustainable mining and resource management** to balance economic needs with environmental preservation.
- **Water security** in a changing climate, as glaciers shrink and rivers shift.
- **Coastal protection** to safeguard cities and biodiversity from rising seas and extreme weather.

Understanding geology is not just about looking at the past. It is about preparing for the future.

15.3 A Call to Action: How Can We Make a Difference

Geology is not an abstract science. It affects where we live, how we farm, what we build, and even how we survive natural disasters. Each one of us has a role to play in preserving India's landscapes, in the following ways:

- Learn about the land you live on, its history, its risks, and how to protect it.
- Support policies that promote sustainable land use and conservation.
- Adopt responsible practices, whether in construction, farming, or energy use, that minimize environmental impact.
- Encourage scientific research and geological awareness to help prepare for future challenges.

15.4 India's Ever-Changing Story

India is a land in motion, always shifting, always reshaping itself. The mountains will continue to rise, the rivers will carve new paths, the land will tremble and heal.

We are only a small part of this grand story, but the choices we make today will determine how this landscape survives for future generations. By respecting and understanding our land, we can ensure that India's geological wonders remain a source of knowledge, beauty, and life for centuries to come.

Further Reading and Resources

For readers who wish to explore India's geology further, the following resources are recommended.

Books

- Geology of India and Burma – M. S. Krishnan. A classic and comprehensive reference on Indian geology.
- The Story of Earth – Robert M. Hazen. An accessible and engaging account of the planet's geological evolution for general readers.
- Earth: An Introduction to Physical Geology – Edward J. Tarbuck and Frederick K. Lutgens. One of the most widely used introductory geology texts, covering plate tectonics, rocks, and natural hazards.
- India's Geological Heritage – Geological Survey of India. A government publication documenting India's key geological sites and their scientific significance.

Online Resources

- Geological Survey of India (www.gsi.gov.in) – The official website of India's premier geological organisation, offering maps, publications, and research reports.
- National Center for Seismology (seismo.gov.in) – Real-time earthquake data, seismic zone maps, and research on India's seismic activity.
- USGS Geology and Geophysics (usgs.gov) – The United States Geological Survey offers world-class educational material on plate tectonics, volcanoes, earthquakes, and minerals.

- NCERT Geography Textbooks (ncert.nic.in) – India's national curriculum textbooks for Classes 6–12 provide excellent introductions to physical geography, soil types, and river systems in the Indian context.

- NCERT Geography Textbooks (ncert.nic.in) – India's national curriculum textbooks for Classes 6–12 provide excellent introductions to physical geography, soil types, and river systems in the Indian context.

Glossary of Key Geological Terms

Alluvial Soil – Fertile soil deposited by rivers, commonly found in the Indo-Gangetic Plain.

Aquifer – Underground layers of rock or sediment that hold water and supply groundwater.

Basalt – A volcanic rock formed from cooled lava, found in the Deccan Plateau.

Biodiversity – The variety of life in an ecosystem, affected by geological changes such as erosion and climate shifts.

Carboniferous Period – A time period (around 300 million years ago) when dense forests existed, leading to the formation of coal deposits in India.

Continental Drift – The slow movement of Earth's continents over millions of years due to plate tectonics.

Crust – The outermost layer of the Earth, where geological activity occurs.

Deccan Traps – A vast volcanic region in central India formed by massive eruptions around 66 million years ago.

Deposition – The process of sediments being laid down by rivers, wind, or glaciers, shaping landscapes like the Indo-Gangetic Plain.

Erosion – The process by which wind, water, and ice wear away rocks and soil over time.

Epicenter – The point on the Earth's surface directly above where an earthquake originates.

Fault Line – A fracture in Earth's crust where movement occurs, causing earthquakes.

Fossil – The preserved remains or traces of ancient plants and animals, found in sedimentary rocks.

Gondwana – An ancient supercontinent that included India, Antarctica, Africa, and Australia before breaking apart.

Glacier – A large, slow-moving mass of ice that shapes landscapes through erosion and sediment transport.

Himalayan Orogeny – The geological process that formed the Himalayas due to the collision of the Indian and Eurasian plates.

Igneous Rock – Rock formed from cooled lava or magma, such as basalt in the Deccan Traps.

Indo-Gangetic Plain – A vast flatland created by the deposition of sediments from the Himalayas over millions of years.

Landslide – The sudden movement of rock and soil down a slope, often triggered by earthquakes or heavy rainfall.

Lava – Molten rock that erupts from a volcano and cools on the surface.

Magma – Molten rock beneath the Earth's surface that forms lava when it erupts.

Metamorphic Rock – Rock that has been changed by heat and pressure, such as marble from limestone.

Orogeny – The process of mountain formation due to tectonic forces, as seen in the Himalayas.

Plate Tectonics – The theory that Earth's crust is divided into large plates that move and interact, shaping the planet's surface.

Precambrian Era – The earliest part of Earth's history, when the first continents and life forms appeared.

Red Soil – Soil rich in iron oxides, giving it a reddish color, commonly found in the Deccan Plateau.

Seismic Zone – A region prone to earthquakes due to tectonic activity, such as the Himalayan region and parts of North India.

Sedimentary Rock – Rock formed from layers of deposited material, often containing fossils.

Subduction Zone – A tectonic boundary where one plate moves beneath another, often creating mountains and earthquakes.

Tectonic Plates – Large sections of Earth's crust that move and interact, causing earthquakes and shaping landforms.

Tsunami – A large ocean wave caused by underwater earthquakes or volcanic eruptions.

Volcanism – The process of magma reaching Earth's surface and forming volcanoes, as seen in the Deccan Traps.

Hotspot – A region of intense volcanic activity caused by a plume of abnormally hot mantle material rising from deep within the Earth. The Deccan Traps are believed to have been formed over a hotspot.

Regur Soil – The black cotton soil found across the Deccan Plateau, formed from weathered volcanic basalt. It is highly moisture-retentive and rich in minerals, making it ideal for growing cotton.

Tethys Sea – The ancient ocean that once existed between the Indian Plate and the Eurasian Plate. When India collided with Asia, the Tethys Sea was squeezed shut and its sediments were thrust upward to form the Himalayas. Marine fossils found in the Himalayas today are remnants of this ancient sea.

Laterite – A soil and rock type rich in iron and aluminium, formed in hot and wet tropical climates. Laterite is found extensively in Kerala, Karnataka, and Goa, and was historically used as a building material.

Palaeochannel – The remnant of an ancient river channel that has since been abandoned and buried by sediment. Palaeochannels beneath the Indo-Gangetic Plain provide evidence of major rivers that once flowed across the subcontinent, including the legendary Sarasvati.

About the author

<hr>

Siva Prasad Bose is an author of introductory guidebooks on aspects of Indian laws. He is currently retired after many years of service as an electrical engineer in Uttar Pradesh Power Corporation Limited. He received his engineering degree from Jadavpur University, Kolkata and has a law degree from Meerut University, Meerut and a BSc from MMH College, Ghaziabad. His interests lie in the fields of family law, civil law, law of contracts, and areas related to conservation and ecology. He lives in Delhi.

Other Books by Siva Prasad Bose

Introduction to Conservation of Wildlife in India

Introduction to Conservation of Indian Monuments

About the Author

Siva Prasad Bose is an electrical engineer by profession. He is currently retired after many years of service in Uttar Pradesh Power Corporation Limited. He received his engineering degree from Jadavpur University, Kolkata and has a law degree from Meerut University, Meerut. His interests lie in the fields of family law, civil law, law of contracts, and any areas of law related to power electricity related issues.

Read more at https://sivaprasadbose.wordpress.com/.